GREAT MINDS OF SCIENCE

Gregor Mendel
Father of Genetics

Roger Klare

Enslow Publishers, Inc.

44 Fadem Road	PO Box 38
Box 699	Aldershot
Springfield, NJ 07081	Hants GU12 6BP
USA	UK

For Connie, Katie, Rachel, Sarah, Arthur,
and McCoy.

Library of Congress Cataloging-in-Publication Data

Klare, Roger.
    Gregor Mendel: father of genetics / Roger Klare.
      p. cm. — (Great minds of science)
    Includes bibliographical references and index.
    Summary: Examines the life and work of the nineteenth-century
Austrian monk who discovered the laws of genetics.
    ISBN 0-89490-789-1
    1. Mendel, Gregor, 1822–1884—Juvenile literature. 2. Geneticists—
Austria—Biography—Juvenile literature. [1. Mendel, Gregor, 1822–1884.
2. Geneticists.] I. Mendel, Gregor, 1822–1884. II. Title. III. Series.
QH31.M45K43 1997
575.1'1'092—dc20
    [B]   96-35791
                                       CIP
                                       AC

Printed in the United States of America

10 9 8 7 6 5 4 3 2

Illustration Credits: Connie Schmittauer, pp. 44, 46; Roger Klare, pp.
8, 18, 23, 26, 35, 39, 50, 61, 65, 69, 73, 79, 90, 92, 97, 105; Stephen
Delisle, pp. 12, 47, 52, 54, 56, 59, 103.

Cover Illustration: Roger Klare (inset); © Valerie Hodgson/Visuals
Unlimited.

Contents

Family Traits

ANTON MENDEL HAD A FINE HOUSE. HE built it himself out of stone and tile. In the early 1800s no one else in his village built houses this way. He was a German living in the village of Heinzendorf in a region called Silesia. (Today the village is called Hynčice, Czech Republic.)

Anton Mendel also had a fine orchard. He planted it with improved types of fruit trees. Anton was a hard worker, like his father and grandfather. His grandfather started as a tenant farmer and saved enough money to buy his own land. Now the land had passed on to Anton. He was a peasant and a farmer, but at least he had his own land. Anton was stout and strong from years of hard labor.

Silesia was part of the very large Hapsburg Empire. As a soldier, Anton spent eight years defending the empire. In his travels he had seen some of the world. But he was more than ready for his soldiering days to end. When the war with Napoleon was over in 1815, it was time to go home and raise a family.

Anton and Rosine Mendel were married in 1818. Rosine came from a family with talent. Her uncle wanted to learn so much that he educated himself. He then became the first regular teacher in the village.

Anton and Rosine named their first child Veronika and their third child Theresia. Johann was their second child. He was born on July 22, 1822. His parents named him after an uncle on Anton's side of the family. Years later, when he became a monk, Johann would take Gregor as his first name. Rosine hoped that Johann might become a teacher like her uncle. If not a teacher, then perhaps he might be a priest. Johann became a teacher and a priest and something else besides. He became a scientist whose work started a new science called genetics.

Genetics is the study of heredity. That's a subject everyone knows a little bit about. People have always known that children look like their parents. One can see heredity at work: features or traits such as eye color pass from parents to offspring. People have always wondered how heredity works. Mendel's work would provide some answers.

Mendel did experiments with garden peas. He looked at seven traits in his pea plants. One trait was height. Some pea plants were tall and others were short. Mendel knew that this trait passed from parent plants to their offspring. He used a tall pea plant as one parent and a short one as the other parent. He wanted to see what the offspring would look like. Would they be tall, short, or somewhere in between? The offspring were all tall, and he wanted to know why.

For eight years he raised his peas and collected his data. When he was done, he began to see

THEODOR CHARLEMONT

Gregor Johann Mendel's work with pea plants started a new science called genetics.

how nature works. He saw that nature plays by certain rules.

Mendel published his work in 1866. In 1900 three scientists used his work to start the brand-new science of genetics.[1]

Others before Mendel had tried to unlock the secrets of nature. What he did was to find the meanings they had missed. Mendel described his results in a new way. He used numbers. At first, people couldn't understand this mixing of biology and numbers. Today, biologists use numbers all the time in their work. However, it was a new idea back then.

What was it about Gregor Johann Mendel that helped him to see what others had missed? He made good use of his family traits. He did not inherit these traits. He learned them as he grew up.

His father taught him the value of hard work. The two of them spent many hours in the orchard and garden. Young Johann learned how to take care of fruit trees and garden plants.

His mother taught him the value of thinking

and planning. She was a quiet and thoughtful woman. Johann was also quiet and thoughtful, though he enjoyed being around people. He learned how to plan carefully and keep good records.

From his family he learned something else. He learned not to give up. This helped him carry out his eight years of experiments with peas. This helped him become head of a large monastery. And this also helped him in his struggle to gain an education.

That struggle was about to begin.

Getting an Education

JOHANN MENDEL'S EDUCATION BEGAN AT home and continued at his first school. At home, he learned about fruit trees and hard work from his father. At school, he learned about science and many other subjects. The school also had its own garden. There Johann learned about plants and beekeeping.

Johann was a talented boy. His teacher Thomas Makitta saw that right away. When Johann finished at his first school, Makitta spoke to Johann's parents. He hoped they would continue the education of one of his best students. In the town of Leipnik was a school that Johann

Johann Gregor Mendel was born in Heinzendorf, in the Habsburg Empire. Today, the village is called Hynčice and is in the Czech Republic. In the 1800s large empires dominated Europe.

could attend. Two boys from the village had just gone on to that school.

During the holidays the two boys returned to town with tales about their school. Johann listened and started to dream about going there. He knew what he wanted to do.

His father, though, had other plans. He wanted to leave his land to his only son. Anton saw that Johann was strong and would make a good farmer. Yet Anton was well aware that a peasant's life was hard. Three days out of each week he had to work without pay for the ruling nobleman.[1]

Despite this, Anton still wanted to pass the land on to Johann. He wanted it to stay in the family. Then Rosine told Anton that education was the only way to escape the hard life of a peasant. At last Anton agreed with her. They would send eleven-year-old Johann to the school at Leipnik. He would be about thirteen miles away from home.

Johann did not let his parents down. He worked hard and was soon at the head of his class. His schooling there lasted just one year, and by then his course was clear. He was going to continue his education. On December 15, 1834, Johann entered Troppau High School. The town of Troppau was even farther away from home. Anton and Rosine were now sure that their son would never be a peasant.

Johann's high school teachers taught him much about science. They were making nature exhibits at a nearby museum. They were also recording weather data, something Johann would later do.

Sending Johann to this school was a good idea. Yet Johann knew how little money his parents had—most of it was spent maintaining the house and property. How would they pay for his schooling?

The solution was to put Johann on "half rations." Rosine would send food packages on the twenty-mile trip to the town of Troppau. It was not enough, and Johann went hungry more than once. Though he couldn't fill his stomach, he did very well at filling his mind. He was one of the best students.

Johann was expanding his mind in other ways. He tried writing to express his thoughts. In a poem he revealed a glimpse of the man he would become. A man who would be both a scientist and a monk. He wrote that science would drive away

> *"The gloomy power of superstition*
> *Which now oppresses [weighs down] the*
> *world."[2]*

In 1838 Johann had to turn away from these thoughts. He had to face a new challenge. That year his father was injured by a rolling tree. Johann quickly went to visit Anton. His father knew that he could no longer work as a farmer. He also knew that he could no longer support Johann.

At the age of sixteen, Johann worried about his father's future and his own. He realized that he must somehow continue his education. Though young, he was starting to develop his problem-solving skills. He quickly came up with a brilliant solution. He took an exam to become a private tutor. As it turned out, Johann was very good at teaching. It would give him enough money to continue his schooling. Little did he know that it would be the beginning of a long teaching career.

That year was a stressful one for Johann. In 1839 he became ill and had to stop his schooling. He stayed at home for several months while he regained his health. Then he returned to school. He had some catching up to do, but nothing could keep him from finishing high school. On August 7, 1840, he graduated with the highest honors.

From one ending came another beginning. Now a young man of eighteen, Johann began to map out his life. His goal was to continue his studies, and for that he must go elsewhere. In the city of Olmütz was a learning center. There he would study science and other subjects.

Money was still a major problem. He began at once to look for work. Johann would later write about this trying time in his life. He preserved his thoughts in a short autobiography. He wrote how he tried to find work "as [a] private tutor in that city, but failed to do so for lack of friends and recommendations."[3]

Again Johann felt the stress of his struggle. He fell ill in 1841 and again spent time resting at home.

Then in August 1841 Johann's fortunes began to change. Because Anton could no longer work, he decided to sell the farm. He would sell it to Alois Sturm, the husband of Johann's older sister Veronika. Alois had been running the farm ever since Anton's accident. Anton wanted to provide for Rosine, Johann, and Johann's younger sister Theresia. The sale of his farm would take care of everyone. Or so he hoped.

Johann knew that his parents always did their best. Yet he knew there just wasn't enough money to pay for his schooling. He still had his dreams. How could he reach them?

At this point Theresia gave him the help he needed. She had planned to use her share of the

Throughout his life, Mendel remained close to his family. Shown here are his sisters Theresia (left) and Veronika (right), and Leopold Schindler, Theresia's husband.

farm sale for marriage. She gave Johann part of that money for his schooling. Things still worked out well for her, for she married and had three children. Johann never forgot his sister's kindness, and he repaid her by taking care of her children. Later he would pay for their schooling. He would be close to them for the rest of his life.

Now full of hope, Johann returned to Olmütz. This time he was successful in finding students to tutor. With his money worries over for now, Johann could pursue his studies. His favorite subject was physics. His teacher Professor F. Franz was both a scientist and a monk. Like others before him, Professor Franz saw that Johann had a mind for science.

This meeting with Professor Franz would be an important turning point in Johann's life.

Becoming a Priest

AT THE AGE OF TWENTY JOHANN FACED one more roadblock in his life. He had just finished a two-year course of study. Now Johann didn't know what to do. He wanted to go to a university, but how could he afford it? He had to think this through. By now he had practice solving problems and overcoming roadblocks.

Johann came up with a plan of action. He decided to speak with Professor Franz. That turned out to be a good plan, for Franz was eager to help. So the two of them sat down to talk about Johann's future.

Professor Franz knew that Johann wanted to

learn more about science. And he knew a way to do it. He suggested that Johann join the monastery in Brünn. Professor Franz himself had lived in the monastery while teaching in that city. Recently a friend at the monastery had written to Professor Franz. He asked him to speak with young men about becoming monks. Professor Franz agreed to recommend the best young men, and now he was talking to one of them.

Johann thought long and hard about what he heard. He would be joining the Order of Saint Augustine. He would follow a code of rules and wear a tunic and cape. What would his life as a monk be like?

Then he remembered Father Schreiber, the priest from his hometown. Father Schreiber taught science and even wrote a book about it. He also knew a lot about fruit trees. He had picked out many young trees from France and had given them to the people at home. Johann knew one person who received some of the young fruit trees—his father.

Johann saw that if he could learn about science, he could put it to good use. So he decided to take up Professor Franz's suggestion. He would join the monastery and would no longer worry about having enough food to eat. Then he would become a priest. In his spare time he would study science.

On July 14, 1843, Professor Franz wrote to his friend at the monastery. He told him all about Johann. In his letter he described Johann as "a young man of very solid character."[1]

On September 7, 1843, the Order of Saint Augustine accepted Johann. Then on October 9, he moved to the monastery and had a new home. That day he also had a new first name. He chose it as his religious name. From now on he would be known as Gregor.

The monastery of Saint Thomas was founded in 1359 and moved to a new location in 1783. Surrounded by small houses, it was at the edge of the city of Brünn. The building had just one floor with an attic and covered a large area. At the rear of the building was a small clock tower

In 1843, Johann Mendel became a monk when he was accepted into the Order of St. Augustine. He would now be known as Gregor Mendel.

above the library. Below was a small garden. It was here that Gregor would do his famous experiments with peas.

Brünn was the capital of a place called Moravia. (Today Brünn is called Brno, and is located in the Czech Republic.) The name of the city meant "hill town." It marked the spot where two rivers flowed together.

Gregor spent those first days getting to know his new home. What a change it was from his old village! Yet in some ways it was similar. It was peaceful and quiet here with plenty of time to think. But was he cutting himself off from the rest of the world in this place? Could he still learn about science here?

Gregor did not need to worry. Cyrill Napp, the abbot, or head of the monastery, was a man of great energy. He was making this place into a center of learning. He invited great scientists and artists to visit the monastery and share their knowledge. Abbot Napp was a good host.

Gregor soon learned that many of his fellow monks were also scientists. They taught science to high school and college students. Gregor could hardly believe his good fortune. Here was his chance to learn firsthand about science from those who knew it. He later wrote that his "fondness for natural science grew with every fresh opportunity for making himself acquainted with it."[2]

Abbot Napp was always looking for monks with a talent for science. Gregor soon caught his eye. Abbot Napp told Gregor about botany lectures at a college nearby. At these lectures Gregor learned how to pollinate plants by hand. He would use what he learned in his experiments with peas.

Abbot Napp was a man of many talents. He wrote a book about growing improved kinds of fruit trees. He helped raise money to fix up the monastery buildings. And he was director of high school education in the area.

Abbot Napp also insisted that all rules be followed. One time he had to remind Gregor about this. Gregor was attending the classes he needed to be a priest. Abbot Napp found out that

Gregor was attending classes without wearing the proper cap on his head. He made it clear that Gregor must wear a cap like the other students. Rules are rules, said Napp. There would be no exceptions.

Gregor attended classes at the Brünn Theological College. He began his studies in 1845 and completed them in 1848. In their reports, his teachers described him as a hard

The head of the Brünn monastery was Abbot Napp, who Gregor Mendel (standing second from right) looked up to and respected. Napp is pictured here in the front row, seated second from the right.

worker with high character. With their blessing, he took the vows to obey the rule of Saint Augustine. By the time he was twenty-five, Gregor achieved what his mother had hoped for. On August 6, 1847, he became an ordained priest. He would now be known as Father Mendel. Due to a shortage of priests, Mendel became one before he finished his studies.

A year later Mendel became assistant pastor at a church in Brünn. Most of the church members were Germans, though a large number were Czechs. Mendel spoke German and always thought of himself as a German. Now he needed to learn the Czech language because he had to speak it from time to time. Mendel was trying to adjust to all the changes in his life. While he was learning the job of a priest, big changes were happening outside the church.

Revolution was in the air. The Hapsburg rulers were losing power and the Czech people were demanding their rights. Before long the uprising was over, and a new government came to power. Again the people had little freedom, but one good thing happened. The forced labor that Mendel's father had to perform was over.

Meanwhile, Mendel struggled with his own job. He had no problem preaching to the church members. His problem was the daily visits he made to the sick and dying. Mendel had always been a nervous person. Now when he saw sick people, Mendel began to feel ill himself. Soon he found it more difficult to carry out all his duties as a priest. He now had another problem to solve. What could he do?

This time, though, he didn't have to solve it all by himself. Abbot Napp had a plan for him.

Becoming a Scientist

ABBOT NAPP'S PLAN WAS TO RELIEVE Mendel of his duties. Napp had sent a letter to the bishop. He stated that Mendel would be much better as a teacher than as a priest. Napp was well aware of all the time Mendel spent learning about science. He felt Mendel could share some of his knowledge with high school students. Since Abbot Napp was head of high school education, he knew where Mendel might fit in.

In September 1849 Mendel received his appointment. He would teach mathematics at the high school in Znaim, a town in southern

Moravia. However, his teaching job was only temporary. To gain a permanent job, Mendel would have to pass an exam for high school teachers.

For now Mendel was eager to start. He had no training in teaching, and of course, did not have a college degree. Yet he did have experience as a tutor. So he began at once to learn his new job. The students and the other teachers liked their new teacher. Mendel's hard work and honesty came out in his teaching.

Sometimes, though, he was a little too honest. One day the bishop came to inspect the high school. Mendel and Bishop Schaffgotsch did not always agree on things. The bishop was also very overweight. On the day of the inspection, Mendel without thinking spoke what was on his mind. Mendel remarked, "He carries about with him more fat than understanding."[1] His words got back to the bishop, who of course was none too pleased. The two of them would continue to have their differences in the years ahead.

The rest of the school year passed quickly.

The other teachers urged Mendel to take the teacher's exam so he could get a permanent job. For Mendel this was a big step. Nearly all teachers took this exam only after years of study at a college. Yet he could take it without college training. After all, he had spent years of his life learning on his own.

So he decided to take the exam for science teachers. He received his test papers on May 10, 1850.

The first question was about the weather. Mendel had to explain all about the air and how winds happen. The second question was about geology. Mendel had to explain the differences between rocks formed by water and by heat.

Mendel had several weeks to complete this written test. In July, Mendel learned that he had passed the first question, but he failed the second. Yet the exam was not over. There would be more tests in August, and for these Mendel had to go to Vienna, capital of the Hapsburg Empire.

Again Mendel had two test questions. For the

first question, Mendel had to explain all he knew about magnetism. For the second, he had to explain all he knew about mammals. Mendel passed the first question. He didn't pass the second one. He really didn't know much about mammals, but he tried to come up with an answer. Mendel then took an oral test of his science knowledge. Mendel failed, and the exam was over.

Mendel was very disappointed. Try as he might, he just couldn't make up for his lack of training. Still he would not give up. He would just have to work harder. He would go on learning on his own if he had to. Mendel went back to the monastery full of plans for his future.

In 1851 Abbot Napp wrote to one of the professors who had tested Mendel. Napp wanted to know why Mendel had failed his exam. The professor, Andreas von Baumgartner, replied that Mendel needed training in science. Baumgartner thought Mendel should go to the University of Vienna. It didn't take much to convince Abbot Napp that this was the best thing to do. He quickly made the plans, and in October Mendel was on his way.

Mendel had to fend for himself, since there were few rooms in the monasteries in Vienna. He did manage to find a room in a religious house. Now he could turn all his attention to science.

Mendel was lucky and he knew it. For the first time in his life, Mendel had the freedom to pursue his favorite subject. He was going to make the most of it. He would always be grateful to Abbot Napp for giving him this chance of a lifetime. Mendel was now twenty-nine years old.

Mendel's first class was in physics. His teacher was Christian Doppler, who was famous for discovering the Doppler effect. A person standing near an approaching train will notice the Doppler effect. As the train approaches, the whistle sounds shrill. As the train passes by, the whistle sounds deep. The whistle sound seems to change. That's because the whistle is moving first toward the person, then away.

Professor Doppler taught Mendel how to do experiments. Mendel would later use some of what he learned in his own experiments. Doppler also taught Mendel more about mathematics and how to use it. This too would be valuable later.

Mendel was starting to feel more comfortable in the world of science. He was starting to think like a scientist. In fact, he was becoming one.

He continued to learn about physics. He came across a book written by an author he instantly remembered—Professor Baumgartner. He had tested Mendel about the weather and magnetism for the teacher's exam. In his book Baumgartner talked about what scientists try to discover and how they do it. Mendel studied the words carefully. He was seeing a whole new way of looking at nature.

Mendel read that nature follows certain rules. The goal of scientists was to discover these rules and explain them. At first, Mendel did not see how to do this. Then he read on. Carefully planned experiments were the key to finding the rules.

During his Christmas break at Brünn,

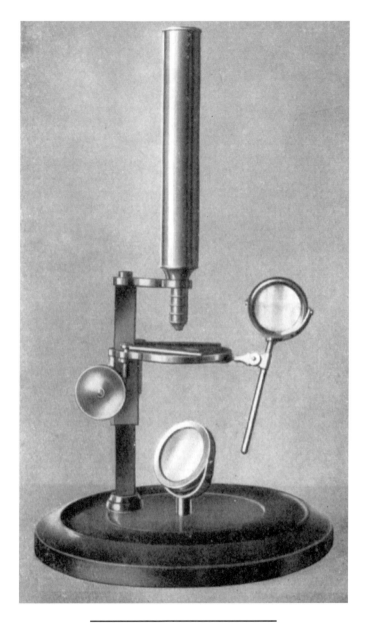

Abbot Napp sent Gregor Mendel to the University of Vienna for training in science. Mendel would use this microscope to help him teach science to students.

Mendel thought about what he learned. He also spent the Easter holiday in 1852 at the monastery. Then it was back to Vienna and the beginning of classes in botany. In these classes Mendel learned more about plants and their parts.

In October 1852 Mendel went back to his hometown. The occasion was the marriage of his younger sister, Theresia. It was a joyous time for Mendel, as he had a few days to visit with all the family members. The days, of course, did not last. Soon he was on his way back to Vienna.

In his next botany class, Mendel was to meet a brilliant teacher. It would be another turning point in his life.

Franz Unger taught Mendel about the structures of plants and how they work. Professor Unger also taught much more. He loved science and passed this love on to his students. He was also not afraid to speak his mind. Unger believed that life gradually evolves, or changes over time. He did not know how life changes, but he was sure that it did. A few years later a man named Charles Darwin would provide some answers. Darwin had a great idea, and in 1859 he published a book about it. The title was *On the Origin of Species by Means of Natural Selection*. But in 1852 most people believed that life stayed the same and did not change. They did not believe in evolution.

Mendel, though, was fascinated. He agreed with Professor Unger that life changes over time. Why, Mendel could see that the earth's surface changes over time. It made sense to think that life did the same.

Mendel enjoyed all that he was learning. He couldn't stop thinking about it. When he took a break from his studies, he had some time for his parents. During the Easter holiday in 1853 he sat down and wrote a short letter. He asked how Theresia and her husband were doing. Mendel described himself as "perfectly well all the time, and hard at work."[2] As he wrote, a huge snowstorm was filling the skies above Brünn. Perhaps it reminded him of the thoughts filling his head.

At the end of July, Mendel returned to the

monastery. He was ready to share what he learned about science with a new group of students. The next year he again found a job as a temporary teacher. Mendel taught physics and natural history at the Brünn Modern School. The school was similar to today's high school. Mendel was back in a job that he liked and was good at. Years later many former students remembered their cheerful teacher who was always helpful and kind.

Mendel wanted to take his teacher's exam a second time. He was ready, he thought. He prepared himself, and in May 1856 it took place. Yet from the start Mendel was in trouble. He started the written part of the test, but the stress got to him. He felt ill. He knew he had to stop. He did not complete the test, but the ordeal at last was over.

Mendel would never take the teacher's exam again. He would remain a temporary teacher at the modern school for the next twelve years.

Mendel was again disappointed, but that soon wore off. He had already started a new

After leaving the University of Vienna, Mendel took a job as a temporary teacher at the Brünn Modern School. In this picture of the school's teachers Mendel is seated, second from the right.

project. It was something big. Professor Unger had given Mendel a problem to solve. Mendel knew that he just had to find the answer. What he found would someday make him famous. Through his work, he would teach a very large class—scientists around the world.

Experiments with Peas

EVER SINCE THE SUMMER OF 1853, MENDEL
had been thinking about what Professor Unger
had taught. Mendel had learned about traits in
plants. He tried some breeding experiments
with ornamental plants. He saw that certain
flower colors would turn up again and again.
Why did this happen? Nature must be following
some rules.

Mendel wanted to find the rules. Right away
he saw how hard this would be. He did, though,
see how he would start. He should find out what
other people had learned about plant breeding.

Plant breeders were always looking for ways

to make changes. The breeders were changing plants to improve them. A fruit-tree grower would take the pollen from one tree and pollinate another tree of the same kind. The grower would try to create a new and better variety of fruit tree. When this worked, the grower could see the results. It seemed mysterious. No one knew why this worked, but it did.

Mendel began reading about scientists who did experiments with plants. He purchased a book by a German botanist named Carl von Gärtner. This book contained the results of nearly 10,000 experiments.

Mendel carefully studied Gärtner's procedure. In one experiment Gärtner started with two varieties of corn. One was tall and had red-striped seeds, and the other was short and had yellow seeds. He used one plant to pollinate the other. The offspring were known as hybrids. Gärtner then looked for parental traits in the hybrids.

Mendel read Gärtner's book again. The words were making more and more sense.

Mendel saw that nature must be following some rules when two plants create hybrids. But what are the rules for hybrids? By doing his own experiments with plant hybrids, he would try to find the answer.

Mendel opened Gärtner's book to the inside of the back cover and began to write. He was making some notes about traits in pea plants. He saw that peas would be good plants to use. He could grow them easily. And he could pollinate the plants himself. Now that he had a plan in mind, he went to see Abbot Napp.

The abbot always enjoyed talking with Mendel. Abbot Napp saw that he had made a good decision to send Mendel for science training. He hoped that Mendel could make use of it. So he was interested when Mendel came to him with a request. Mendel needed some space to grow some peas. Abbot Napp said there was some garden space he could use. It was not very big, but Mendel thought that was fine. It would be enough for now.

In 1854 Mendel began work in the garden. A

plan for some experiments was forming in his mind. During that year and the next, he planted different varieties of the garden pea. He looked at many different features or traits in the plants and seeds. He wanted traits he could see easily.

At last he chose seven traits. His choices were:

1. Seed shape—round or wrinkled
2. Seed color—yellow or green
3. Plant height—6 to 7 feet or 9 to 18 inches
4. Flower position—along the stem or at the end of the stem
5. Color of unripe pods—green or yellow
6. Pod shape—smooth or ridged
7. Color of seed coat—gray-brown or white

In the spring of 1856 Mendel began his experiments with peas. He would study two forms of each trait. He wanted to see how a trait would pass from parent plants to their offspring. Would one form of the trait mask the other? Or would the two forms mix and blend?

To find out he knew that he would need many

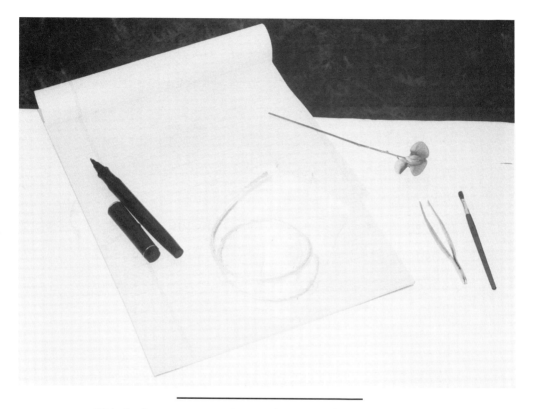

This is the necessary equipment for cross-pollinating pea flowers. Mendel used items like these to perform his work.

varieties of garden peas. One produced only round seeds. The other variety produced only wrinkled seeds. He took pollen from some of the peas that produced round seeds. He then pollinated by hand some of the peas that produced wrinkled seeds. He tried this the other way around with the rest of the plants. He took pollen from peas with wrinkled seeds and pollinated the peas with round seeds.

He slowly went from flower to flower. Inside each flower he found the stamens, the male part of the flower. The stamens produce the pollen Mendel collected with a camel-hair pencil. He also found the pistil, the female part of the flower. Inside the pistil is an ovary with eight to thirteen ovules. Each ovule contains an egg cell.

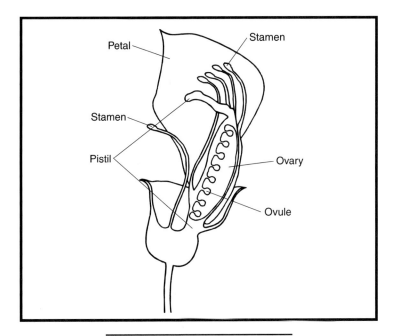

These are the parts of a pea flower. Mendel collected pollen from the stamens of one plant and placed it on top of the pistil of another plant. A single pollen grain would then reach the ovule.

Mendel carefully placed the pollen from one plant on the pistil of another plant. From the top of the pistil, one pollen grain would form a pollen tube and reach an ovule. When that happened a sperm cell from the pollen grain would fertilize the egg cell. The plant would then make a new pea seed.

It was slow and tiring work. However, Mendel was patient. He later wrote a letter to a scientist and explained what kept him going. He wrote: "My happiness will be redoubled [increased] if, by my experiments, I can only succeed in finding the solution of the problem."[3]

After Mendel finished working on a pea flower, he wrapped that flower with a paper bag. He wanted to keep out a bee or other insect. An insect might bring pollen from another pea flower. That could ruin his experiments. Mendel decided that it was better to prevent any problem from happening. Because he pollinated all the flowers himself, Mendel knew where the pollen had come from.

Pea plants grow fast, so soon Mendel had

some results. During autumn he opened the ripened pea pods to look at his hybrid seeds. They were all round. He found no wrinkled seeds—that form had somehow disappeared. Now Mendel had to find out what was going on.

The next year Mendel planted all the round hybrid seeds. The hybrid seeds quickly grew into hybrid plants. Before he knew it, the pea flowers appeared. This time, though, he tried something different. Before, he had taken pollen from one flower and placed it on another flower. That was cross-pollination. This time he let the peas pollinate without his help. Pollen from one flower would land on the pistil from the same flower. This is self-pollination.

Before long the plants produced a new crop of seeds. Eagerly he opened the ripe pods. Some of them contained wrinkled seeds. So that form had not disappeared after all! The round form had simply masked the wrinkled form.

Mendel came up with a way to describe what happened. Round stood out, so he called it dominating. (Today we use the word *dominant*.)

This is a pea plant. Pea pods ripen in the autumn, and that is when Mendel was able to see the results of his experiments.

Wrinkled stayed in the background, so he called it recessive. Mendel also clearly saw that something else was going on. The seeds were either round or wrinkled. The two forms had neither blended nor been lost in the hybrid plants.

Mendel then counted the number of round and wrinkled seeds. He had 5,474 round seeds and 1,850 wrinkled ones. Mendel quickly figured out that he had almost three times as many round as wrinkled seeds. He wrote the numbers in his notebook.

Mendel also did an experiment with the trait of seed color. He found that yellow was dominant and green was recessive. He ended up with about three times as many yellow seeds as green seeds. This matched what he found with seed shape. So there had to be some reason for this 3 to 1 ratio. He started to think about it.

He went on to test the other traits. The table shows what he found for all seven traits. For each trait, the results were the same. He always had about three times as many dominants as recessives.

Trait	Dominant Form	Recessive Form
Seed shape	Round	Wrinkled
Seed color	Yellow	Green
Plant height	Tall	Short
Flower position	Along the stem	At end of stem
Unripe pod color	Green	Yellow
Pod shape	Smooth	Ridged
Seed-coat color	Gray-brown	White

Mendel did experiments with seven pea traits.

One year after Mendel began his experiments, his father died. Gregor had been a devoted son. He stepped up his efforts to aid his mother, who was in poor health. He kept in touch with his sister Theresia's husband. In a letter Gregor told about his plans for his mother. He wrote: "I will do everything in my power to ensure that she shall lack for nothing."[4] He sent money and helped make his mother more comfortable until she died in 1862.

Mendel knew that he must go on with his experiments. His parents, though they had known little about them, would have wanted it this way.

6

Discovery

MENDEL WAS NOT FINISHED WITH THE seed-shape trait. His hybrid plants had produced about three times as many round seeds as wrinkled seeds. That was the 3 to 1 ratio. Now he planted these new seeds to see what would happen.

The wrinkled seeds grew into plants that made only wrinkled seeds. But the round seeds were different. One-third grew into plants that made only round seeds. Two-thirds grew into hybrid plants—ones that made both round and wrinkled seeds.

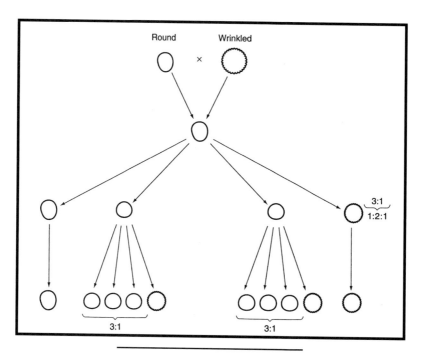

Mendel discovered the classic illustration of the 1:2:1 ratio in his studies.

Mendel saw that he did not have a 3 to 1 ratio. What he really had was a 1 to 2 to 1 ratio.

At last he had made a discovery. It was his first rule or law for hybrids. In science, a law shows how something in nature works.

Mendel didn't give a name to the law he discovered. He could have called it the 1 to 2 to 1 law. It was a law because this was how hybrids behaved. In his other experiments with peas,

Mendel found the same 1 to 2 to 1 ratio. So he was sure. What was true for seed shape was true for the other traits.

Now Mendel wanted to explain his results. He went back to the start of his first experiment.

He had two forms for seed shape—round and wrinkled. One parent plant supplied the round form. The other parent plant supplied the wrinkled form. Writing in his notebook, he used letters in place of the two forms. For round, he used the large letter A. For wrinkled, he used the small letter a.

Mendel thought about his hybrid seeds. One form came from the egg cell. The other came from the pollen. He wrote this combination as Aa or aA. All of these hybrid seeds were round.

The hybrid seeds became hybrid plants. What would happen when these plants made pollen and egg cells?

Now Mendel came up with a great idea. He thought there were two kinds of pollen and two kinds of egg cells. Half the pollen would carry A, and the other half would carry a. Likewise, half the egg cells would carry *A*, and the other half would carry *a*.

If the hybrid plants self-pollinated, what would happen? This table shows all the seed combinations:

Pollen

		A	a
Egg cells	**A**	**AA**	**Aa**
	a	**aA**	**aa**

A seed with *AA* is round. Since round is dominant, it will mask wrinkled when the two are together. So a seed with *Aa* is round. So is a seed with *aA*. That makes a total of three round seeds. The last combination is a seed with *aa*. That one is wrinkled, since wrinkled is all by itself. That's why Mendel had the 3 to 1 ratio.

The *AA* seed will become a plant that makes only round seeds. The *Aa* and *aA* hybrid seeds will become hybrid plants that make both round and wrinkled seeds. The *aa* seed will become a plant that makes only wrinkled seeds. That's why the 1 to 2 to 1 ratio turned up.

Soon Mendel was back in his garden. Though he worked alone, he found someone to talk to about his work. Alexander Makowsky was a fellow teacher at the Brünn Modern School. Trained as a scientist, he was very interested in Mendel's work. Starting in the year 1860, Mendel and Makowsky spoke often with each other. They shared what they were learning.

Mendel was now ready for another experiment. He had seen what happened when he looked at one trait. Now he wanted to look at two traits. This would be harder work because there would be more to keep track of. But he just had to know what would happen.

As before, Mendel started with two varieties of peas. One produced only round yellow seeds. The other variety produced only wrinkled green seeds. Mendel cross-pollinated the two varieties. These pea plants produced hybrid seeds that were all round and yellow. This made sense, since round and yellow are both dominant forms.

The hybrid seeds soon grew into hybrid plants. Mendel then allowed the plants to self-pollinate. He could hardly wait for the plants to produce seeds.

When the pods were ready, Mendel opened them and counted the seeds. He had 556 of them. He wrote down what he saw:

1. Round and yellow—315
2. Round and green—108
3. Wrinkled and yellow—101
4. Wrinkled and green—32

Mendel found seeds just like those from the parent plants. He also found two kinds of seeds that were different. Round could go with either yellow or green. Likewise, wrinkled could go with yellow or green.

Mendel saw what was going on. Four combinations were possible, and he had them all. Mendel had just made a second discovery about hybrids. He was looking at a series of combinations. This was the combination series law.[1]

Mendel went on to test three traits. His results showed that the combination series law was again at work.

In 1863 Mendel finished his experiments

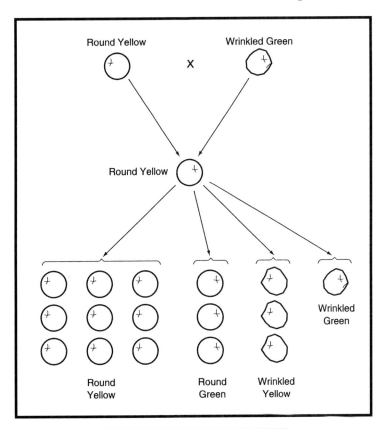

Mendel cross-pollinated peas that made only round yellow seeds with peas that made only wrinkled green seeds. The hybrid plants produced seeds with four combinations of seed shape and seed color.

with peas. It had been long and tiring work. But it was well worth it—he had discovered two laws for hybrid plants. What Mendel had found would be a great help to plant breeders. He wanted to tell others about his work, and he knew where he would start.

Two years before, Mendel and Alexander Makowsky met with a group of scientists. They formed the Brünn Natural Science Society. The members met with each other and shared what they knew about science. Some of them, Mendel recalled, had already been talking about hybrid plants. Mendel saw that this was a good place to start spreading the word about his work.

Mendel gathered all his notes. Then on February 8, 1865, he was ready to talk about what he had learned. About forty members of the natural science society gathered to hear him speak. Mendel was an impressive speaker. Wearing a large hat and black coat, he caught everyone's attention.

The daily newspaper of Brünn covered the meeting. The reporter wrote that Mendel gave a

Mendel knew that his experiments with peas would be a great help to plant breeders. He spoke in front of the Brünn Natural Science Society in an effort to explain his work to others.

long talk about his work with plant hybrids. Mendel spoke about his pea plants and showed how he got his results. The reporter said that the audience enjoyed Mendel's talk.

One month later Mendel again spoke to the society. Again the newspaper covered the meeting. Mendel explained some more about his experiments and how he did them. The newspaper reporter said that Mendel would continue with his work.

All seemed to be going well. Everyone listened to him. However, Mendel didn't give enough information about his work. So no one really understood his discoveries. And no one said they would try to do experiments like the ones he had done.

Still the Brünn Natural Science Society wanted to publish Mendel's lecture. Mendel's title for his work was "Experiments on Plant Hybrids." His paper was published in 1866.

Copies of Mendel's paper would go to universities and science societies all over the world. It would travel to Germany, Great Britain, France, Sweden, Italy, Russia, and the United States. Yet nearly everyone ignored it. For the next thirty-four years, Mendel's paper would remain mostly unknown.

Experiments with Other Plants

MENDEL DID NOT HAVE MUCH TIME TO think about his paper. He was very busy in 1866. He had started experiments with other plants— nor was that all. During that year he had something else taking up his time. Mendel had to deal with the Prussian invasion.

Prussia was the center of the German Empire. During the summer of 1866, soldiers from Prussia invaded parts of the Hapsburg Empire. Mendel's monastery was right in the path of the invasion.

On July 12, about 5,000 Prussian soldiers entered Brünn. Mendel's monastery had to lodge one hundred of them. Luckily, the soldiers didn't stay very long in Brünn. The towns outside the city suffered the most. Mendel told about the invasion in a letter to his sister Theresia's husband. He wrote: "Worse still, the Prussians brought cholera with them, and this terrible illness has wrought havoc [caused damage] among us for at least six weeks."[1] Some time passed before Mendel could get back to his work.

Mendel was ready to continue his experiments with other plants. He had a good reason for wanting to go on. Using peas, he had discovered two laws for hybrid plants. Now he wanted to see if these laws worked on other plants.

Before 1866 Mendel had finished experiments with two different kinds of bean plants. Mendel used beans that grew to different heights—tall or short—just like his pea plants. The beans had different pod shapes—smooth or ridged—also like his pea plants. The beans also had different pod colors—green or yellow— again like his pea plants.

Mendel compared his results with what he found for pea plants. He had tested three traits at a time in peas. He did the same with beans. He figured out all the combinations of traits, and all of them turned up. So the combination series law was again at work.

He also tested the trait of flower color in beans. He was only partly successful, though. Mendel could not decide if flower color followed

Mendel began experiments with beans so that he could compare his results for bean plants with the results of his experiments with peas.

the combination series law. What he didn't know was that many things control flower color in beans and other plants.

Mendel couldn't stop now. He had to do more, but what plants should he test? Mendel knew that now he would need some help from a professional scientist. He had someone in mind.

Mendel had heard about a famous botanist who lived in Munich, Germany. His name was Carl von Nägeli. He was very interested in hybrid plants. On New Year's Eve in 1866, Mendel sent a letter along with a copy of his paper on peas.

This was the first of a series of letters exchanged between Mendel and Carl von Nägeli.[2] In his first letter Mendel described his experiments with peas. Mendel said that he needed to do more work with other plants. One of the many he picked was hawkweed, a plant whose flowers look like dandelion flowers. He asked for advice on working with this plant.

Mendel couldn't have chosen a better expert. As it turned out, Carl von Nägeli was also working with hawkweeds. He wanted to get hybrids of these plants. Nägeli wrote back, saying that he liked Mendel's plans for hawkweeds. He gave Mendel some ideas for further experiments.

Nägeli also gave his opinion of Mendel's experiments with peas. He thought Mendel needed to do more work, and he did not agree with Mendel's explanations.

Mendel, though, was not discouraged. Far from it. He wrote back in April 1867. Mendel expected that Nägeli would disagree with him about the pea experiments. Had he been in Nägeli's place, he would have had the same reaction. Mendel then went on to write some more about his work. He tried to convince Nägeli that he had found some important laws for hybrids. Mendel did not give up easily, of course.

In 1870 Mendel wrote to Nägeli with some more news. Mendel had worked with corn plants, using the trait of seed color. He wrote: "Their hybrids behaved exactly like those of Pisum [peas]."[3] Mendel's laws for hybrids were working with still another plant.

Besides beans and corn, Mendel did experiments with many other plants. He tried several flowers, including one called columbine. In all, he worked with seventeen kinds of plants.[4]

Mendel, however, spent most of his time on the hawkweed plant. He would spend six years trying to discover its secrets. His work with hawkweed was just as important as his work with peas had been. Mendel worked long and hard, but he did not have much success. Mendel had many problems with hawkweed, and he should have chosen something else.

To begin with, hawkweeds were hard to find. Very few of them grew nearby. And to make matters worse, Mendel was becoming overweight in 1867. He had gained so much weight that he was having trouble getting around.

Nägeli came to Mendel's rescue. He supplied Mendel with the seeds from many kinds of hawkweeds. He encouraged Mendel to start new experiments. That made Mendel try even harder to please him. As he had done all his life, Mendel was not about to give up.

Mendel did experiments with seventeen kinds of plants. Shown here is columbine, one of the many plants he worked with.

Mendel had a hard time with his experiments. Hawkweed flowers are very small. They are also hard to pollinate by hand. Mendel did come up with a solution for this problem, however. Using a magnifying glass and mirror, Mendel learned how to pollinate the tiny hawkweed flowers. Yet this work was hard on his eyes. The eyestrain became so bad that by June 1869, Mendel had to stop his work. He continued to have eyestrain and so for about a year did not do any more work. Then he was back at it again.

Mendel had faced a big challenge ever since he starting working with hawkweeds. He had a problem getting any hybrid seeds at all. He cross-pollinated many kinds of hawkweeds. Most of the time he ended up with no seeds. At last, though, he had some success. He had a few hybrid seeds and planted them to get hybrid plants.

To Mendel's surprise, hawkweeds didn't behave like peas. They were not following the 1 to 2 to 1 law.

Mendel did not know that hawkweeds have

something unusual about them. They can produce seeds without needing pollen. The seeds are made directly from the egg cells in the plant. Nothing else is needed. When these seeds sprout, they will all be just like the plant that made them.

Neither Mendel nor Nägeli knew this about hawkweeds. Mendel thought he could figure out what was going on if he kept at it. He knew that could take awhile. So he decided to report on what he had learned so far. In June 1869 he spoke about his hawkweed experiments at the Brünn Natural Science Society. He hoped that he could give another report the next year.

In 1870 the Natural Science Society published Mendel's report. The title for his paper was "On Hawkweed-Hybrids Obtained by Artificial Fertilization." Mendel kept on with his work with hawkweeds, but only for another year. Little did he know that his experiments with plants were about to end. Ever since 1868 Mendel's life had been changing in many ways. Mendel was about to run out of time.

Monastery Duties

IN 1867 ABBOT NAPP DIED. HE HAD BEEN head of the monastery for the last forty-three years. Abbot Napp had done a fine job. He had also been a good friend and supporter of Mendel's work. It was very hard for Mendel to say goodbye.

Now it was the task of the monks to look for a new abbot. Each time the monks chose a new abbot, the government would collect a tax. So it was a good idea to elect someone who would live for a long time. Mendel was forty-five years old.

The monks wanted someone who could get along with everyone. They all liked Mendel. He

Abbot Napp died in 1867. Then, the monks selected Gregor Mendel to be the new abbot at the Brünn monastery.

was known for his thoughtful manner. Mendel also had another thing going for him. The emperor had to approve the election, and he wanted a German abbot. Mendel was a German, while many of his fellow monks were Czechs. In 1868 Mendel became the new abbot. Nearly all the monks had voted for him.

In a letter to Carl von Nägeli, Mendel told him the news. In Mendel's words, "On March 30th the chapter of the monastery to which I belong elected me their lifelong chief."[1]

Mendel thought he could still go on with his experiments. But it was not to be. His work with plants would end three years later.

For now, though, Mendel had many things on his mind. His first task was to face his own doubts. How could he fill the shoes of the late Abbot Napp? That was a very tall order. Mendel had carefully watched the way Abbot Napp had carried out his duties. Mendel decided to keep some things the same and do other things differently. He could not be another Abbot Napp—all he could do was to be Abbot Mendel.

Abbot Napp had tried to get the best men he could find for the monastery. Mendel wanted to do the same. Abbot Napp had also done a good job with the finances of the monastery. Mendel would continue the good job Napp had done.

Mendel saw something he would change. He wanted the church services to look better, so he ordered new robes. Mendel saw something else that bothered him. He noticed that some monks were not living up to the rule of Saint Augustine. This was important, since people saw the monks as model citizens. Yet Mendel was always gentle when he had to correct someone.

In the next few years Mendel would have many changes in his own life. One change happened right away. It was time to say goodbye to his teaching career. He had taught at the Brünn Modern School for the last fourteen years. Though he had been a temporary teacher all this time, he was part of the teaching staff.

His students, of course, loved him. Saying goodbye to them would be hardest of all. Mendel decided to have the school principal announce his departure. That way Mendel could avoid being too emotional about his leaving. He also decided to give his last month's teaching pay to the three poorest boys in his class. They never forgot this act of kindness. Nor did the other students forget their great teacher. Many looked him up at the monastery and spoke with him.

Mendel had moved into new quarters. He had a group of rooms, all with beautiful furniture and paintings. The monastery that Mendel now headed was not short of money. Mendel's new job paid him well, but he was not seeking wealth. He would use his money to help many people with their needs.

He could help his younger sister's three sons. All three went to the Brünn High School at his expense. Two went on to medical school, and Mendel again paid for their schooling. Mendel had always wanted to pay his sister back for helping him with his own schooling. Now he could do that and more.

Mendel gave money to other people in need. In his hometown many people had lost their homes to fires. Mendel helped the residents of the town start a fire station. The grateful residents made him an honorary member of the station.

Mendel also gave of his time freely. He served on many committees and attended many meetings. The same year he became abbot, Mendel was elected vice president of the Brünn Natural Science Society. Yet soon Mendel had little time to attend meetings of the science society. For in 1870 he was elected to the Agricultural Society of Moravia. This group did many tasks, and Mendel took part in all of them.

The agricultural society was in charge of government aid to farmers. Mendel spent time figuring out how to best do this. He also wrote for the society's magazine. In addition, people always wanted to ask him questions about growing fruit trees. Abbot Mendel was in great demand.

As if all this didn't take up most of his time, Mendel still had an important job to do. He was in charge of all the monastery property. That meant he had to visit and inspect all the buildings. Since the monastery had a lot of property, that took up much of his time.

In the spring of 1868 Mendel found out what could happen when he was away. Carl von Nägeli had sent him some hawkweed plants. They arrived just as Mendel set off on a tour. When Mendel got home, the plants were not in good shape. In a letter to Nägeli, Mendel told him what happened. He wrote: "When I got home a few days ago I found, to my great concern, that about half of the potted specimens were dead. . . ."[2] With his monastery duties, Mendel had less and less time for his plants.

Mendel had other trips that took him away from home. He traveled to Rome and Vatican City, the home of the pope. Mendel also went to Berlin and Vienna. When Mendel had any free time, he and Alexander Makowsky went to the Alps.

In November 1873 Mendel sent his last letter to Carl von Nägeli. Mendel said that since 1871 he had ended his experiments with plants. He

Gregor Mendel had many more responsibilities now that he was appointed abbot, and less time to work with his plants. One of the plants Mendel had been trying to work with was hawkweed (pictured above).

decided to send Nägeli all the hawkweeds that he had worked on, about 235 plants.

Mendel had sent Nägeli many letters. Nägeli was one scientist that Mendel thought would understand his work with peas. Yet Nägeli was much more interested in hawkweeds than peas. He also rejected Mendel's laws for hybrids. He still didn't think Mendel had the right explanations.

Mendel had ended his plant experiments, but his science work was not over. He still had time for the weather.

Chasing the Wind

LONG BEFORE HE BECAME ABBOT, MENDEL was a weather watcher. One of his high school teachers had kept a record of the weather. Mendel could trace his interest back to those days. He would later keep his own records of weather data.

A weather watcher had been serving Brünn since 1848. His name was Dr. Pavel Olexik. Mendel got to know Dr. Olexik, and soon the two were working as a team. In 1856 Dr. Olexik asked Mendel to look at past weather records. That was the same year that Mendel began his work with peas. So Mendel was starting two science careers at once—experimenting with plants and watching the weather.

Mendel studied weather records as far back as 1848. By 1862 he had fifteen years of records to look at. He looked at temperature, air pressure, air moisture, wind speed, rainfall, and snowfall. Mendel compared all of these things for each of the fifteen years. He wanted to see the big picture, so he made graphs and charts. Now he could see at a glance how the weather changed over the years.

In 1862 Mendel brought his results to a meeting of the Brünn Natural Science Society. The society published his survey of the weather data in 1863. For the next four years, the society would publish weather data prepared by Mendel. Within a few years people heard of Mendel's work. He even received a letter from a famous Dutch scientist asking about his weather data.

Mendel's work also helped him make a friendship. The secretary of the science society was Gustav von Niessl. The son of an army officer, Niessl had studied many things. Like Mendel, he had studied both plants and the weather. Also like Mendel, he was a teacher. Mendel and Niessl became good friends, and they often discussed plant breeding and the weather.

Mendel wanted to put his weather data to good use. In 1877 he wrote a paper about the value of forecasting for farmers. People had begun using telegraphs to send weather data from place to place. Mendel saw that forecasts for all of Europe would someday be possible. In 1879 Mendel wrote a paper titled "The Foundation for Weather Forecasting." In it he pointed out that many weather stations would need to be built. Europe had only twenty to thirty stations, while the United States had more than one hundred.

In 1878 Dr. Olexik was too ill to be a weather watcher any longer. Mendel took over the job. He used a barometer, a thermometer, and a rain gauge. Three times each day Mendel recorded the weather at the monastery. Then he would send his data to Vienna, where it was combined with other weather data.

Mendel was keeping track of what nature was doing. He felt as if he was chasing the wind. One time, though, the wind was chasing him. It was on that day that Mendel had a brush with death. He was lucky; the monastery church was not. For that was the day that a tornado struck Brünn. Mendel was in his room when the whirlwind struck on October 13, 1870, at two o'clock in the afternoon. Despite his fear, Mendel kept his wits about him and described what he saw and felt. He was used to thinking as a scientist.

The day had started like any day in early fall. Puffy clouds came drifting by. Nothing, it seemed, could go wrong on a day like this. At one o'clock, it was raining north of the city. An hour later the wind struck with no warning. The sky had suddenly turned dark, and within minutes the building began to shake. Doors flew open, heavy desks moved, and plaster rained down. Windows shattered and some roof tiles came flying into Mendel's room. Mendel had only read about these storms, but now he was facing one up close. A tornado was passing overhead!

It seemed as if it lasted minutes. Yet it was over in seconds. Looking out a window, Mendel saw the whirlwind—a black form shaped like an hourglass. The upper part hung from the clouds. The lower part touched the ground. He stared at the strange sight. The shape was not what he thought it would look like. The whole thing acted like smoke rising from a chimney on a damp, calm day.

As the whirlwind moved away, Mendel quickly guessed its size. He thought the lower part was 750 feet high. He thought the upper part was 1,000 feet off the ground. Mendel noticed something else about the twister. It was spinning clockwise. That was odd, since most twisters in the northern half of the earth spin the other way.

Days later, Mendel had time to think more about these storms. What could cause the air to spin so fast? Why does a tornado form?

He thought the cause could be a collision of

two air streams. They would be coming from opposite directions. Based on what people knew, Mendel's idea made sense at the time. (One hundred years later, scientists using computers found that Mendel's idea is not correct. Yet even today they do not know exactly why tornadoes form.)[1]

Four weeks after the storm Mendel spoke about it at the Brünn Natural Science Society. He then wrote a paper titled "The Whirlwind of October 13, 1870." His first few words summed up his feelings. He described a tornado as "extremely disagreeable and dangerous for all those who come into close contact with it."[2]

Few people ever saw Mendel's paper about his tornado. Mendel, though, would never forget the whirlwind. He did not know of any injuries or deaths caused by the storm. He did know about damage to trees and houses. His own church was hit hard, with over 1,300 panes of glass broken. His room was a mess.

Mendel prayed that he would never be that close to a whirlwind ever again.

The Last Years

IN 1874 MENDEL BEGAN THE FINAL struggle of his life. It would take up most of his last ten years. This would be unlike his past struggles to get an education and find a career. This time Mendel would take on the government. His battle was with what he thought was an unjust law.

In that year the Hapsburg government passed a law ordering all monasteries to pay a tax. The purpose of the tax seemed to be a good one. The money would increase the salaries of priests. Yet the tax was steep—about 10 percent of what a monastery was worth. The law would take effect in 1875.

Mendel was not against the idea of increasing the pay of priests. He just thought the law was unjust. So in place of the tax amount due, he sent a smaller amount.

The government quickly answered. Mendel received his money back with a note demanding the full amount due. Still the government was willing to listen. They would lower the tax if Mendel could find a mistake. The government might have thought the monastery property was worth more than it really was. They gave Mendel a way out of the mess.

But he would have none of it. Mendel took a stand and refused to budge. The law was unjust and that was final.

Mendel was not going to give up easily—it was a trait he learned as a boy. His persistence had worked well in his scientific work and in the rest of his life. But it didn't work well in this struggle. The government was not about to give in to this stubborn abbot. In 1876 the government took over part of the monastery property. The government was going to get its tax and that was that.

Facing the tornado just might have been easier for Mendel than facing the government. At least the fury of the storm was soon over; not so the government. Mendel sent letter after letter of protest to the state officials. It was as if he had written to a brick wall. They were not going to change their minds.

There seemed to be no way out for either side. At this point Mendel was appointed vice president of the Moravian Mortgage Bank. The government itself had given the job to Mendel. The idea was that Mendel would now have more money. However, he decided to use it to help others in need and not pay the tax.

On days when he was not fighting with the government, Mendel spent time with his hobbies. He loved to work with flowers, vegetables, bees, and fruit trees. He fondly recalled his days as a young boy when he first learned about gardening. He still loved to be around plants in his spare moments. He had a lot more space to grow plants than he had before. And it was his interest in flowers that helped lead him to experiments with peas.

One day in 1878 a visitor stopped by to see Mendel. The visitor's name was C. W. Eichling, and he came from a French seed-producing company. He wanted to learn about Mendel's experiments with peas. With great delight, Mendel took him around to his garden. The peas were in full flower. Mendel explained that he was working to improve the pea seeds. Mr. Eichling then asked Mendel how he did this work. Mendel replied: "It is just a little trick, but there is a long story connected with it which would take too long to tell."[1]

Another long story was Mendel's love of bees. They had always been a part of the family farm. Though beekeeping was a hobby, Mendel learned as much as he could about it. He kept careful records of all his bees.

Mendel wanted to breed bees, just as he had bred plants. He wanted to see if he could create

One of Mendel's hobbies was beekeeping. Bees were part of his family farm growing up, and the beehives he kept in the garden at the Brünn monastery are shown here.

new varieties. His plan was to breed native bees with those from other lands. To do this, he designed a special hive and placed wire screen over it to separate the bees. He spent much time working on this project.

But breeding bees turned out to be more difficult than breeding plants. His special hive didn't work the way he thought it would. Yet he learned a lot from trying. Also something good did come out of it. A hillside near the monastery was bare and ugly. Mendel made sure that all kinds of flowers were planted on the hillside. It was a great success. The flowers were lovely, and now Mendel's bees would have a ready supply of nectar.

In the last years of his life, Mendel spent time with his fruit trees. He bred new varieties of apples and pears. For this he received a medal from a gardening society in Vienna in 1882.

The next year Mendel wanted to get branches from some special fruit trees. He wrote to Alois Sturm, who owned the farm that once belonged to Mendel's father Anton. Mendel wanted to grow the fruit trees his father had grown. They had always been a part of Mendel's memories.

By 1883 Mendel had struggled with the government for nine years. His protests had gotten him nowhere. The government still had control of part of the monastery property. The government was still collecting the tax. Now Mendel was becoming more isolated at the monastery. Some members of the church were growing tired of his protests. And to make matters worse, Mendel's health was beginning to fail.

He had developed Bright's disease, an illness

In 1882, Mendel received a medal from a gardening society in Vienna because of his work with fruit trees. This is an example of a pear tree.

of the kidneys. His heart was also not well. Mendel was now under a doctor's care.

During his illness, Mendel's closest friends were his two nephews, Alois and Ferdinand Schindler. Both were now medical students. They spent their free time with Mendel and enjoyed talking with him. All three were very good chess players, and they spent many hours playing.

Mendel remained a scientist until the very end. On January 4, 1884, Mendel was writing out his last weather records. He died two days later on Sunday, January 6, 1884. His death was caused by kidney and heart failure. He was sixty-one years old.

The funeral ceremony took place on January 9, at the monastery church. Many attended the ceremony. They were teachers, former students, members of the church, and even members of the government. Also attending were members of the science socicties Mendel served in. The Heinzendorf fire station was there, as were the many people who had received aid and support from Abbot Mendel. He was buried at the Brünn central cemetery.

Some people knew Mendel as a weather watcher. Others knew him as a plant breeder. Yet no one knew that Gregor Mendel's work with peas would become the foundation of a new science. That would happen a few years later.

Rediscovery

WHEN MENDEL'S PAPER ON PEAS APPEARED in 1866, just a few people read it. The paper mostly sat on bookshelves. In 1880 a scientist named Wilhelm Focke wrote a book about hybrid plants. He mentioned Mendel's work many times in his book. Focke's book soon caught the eye of one famous person.

That person was Charles Darwin. He received a copy of the book and looked at parts of it. Yet he did not read about Mendel's work. He missed all those places in Focke's book! No one will ever know what he might have thought about Mendel's work. He would never hear about it.

Mendel, though, had read all about Darwin's work. Darwin wrote a book showing that life changes over time. Mendel had a German translation of Darwin's book and made many notes in it. Mendel agreed with Darwin. Mendel, who did not speak English, never tried to meet with Darwin or write to him.

For the next nineteen years Mendel's work was still mostly unknown. Then two scientists read Focke's book. Their names were Carl Correns and Erich von Tschermak. Both read about Mendel's work, and both saw how important it was. They decided to do something about what they read. They, along with another scientist named Hugo de Vries, told the world about Gregor Mendel. In 1900 they announced their rediscovery of Mendel's work.

The world of science had changed a lot between 1866 and 1900. It was not the same world that Mendel had known. With their microscopes, scientists were learning some of the

In 1900, scientists rediscovered Mendel's work and saw that it could be used as a basis for studying heredity.

secrets of cells. They learned how to stain cells so they could see them. A scientist named Walther Flemming then made an important find. He saw rod-shaped structures called chromosomes.

Scientists started thinking about the chromosomes. What are they doing in cells? Could they have something to do with heredity? Maybe they carry something that tells a pea plant to make round seeds. It would be passed from parent to offspring. Many years later this idea turned out to be right. The "something" passed from parent to offspring became known as a gene.

Meanwhile, the world learned about Mendel's work. Mendel had thought his work would help start a science about hybrid plants. But it didn't turn out that way. In 1900 scientists looked at Mendel's work in a different way. For them, Mendel's work would start a science about heredity. This new science would become known as genetics.

It did make some sense. After all, Mendel showed how traits pass from parents to offspring. That's the basic idea of heredity. The scientists hoped to find more answers, so they looked closely at Mendel's paper on peas. They were looking for laws.

They saw Mendel's ratio of 1 to 2 to 1. This was Mendel's first law for hybrids. He had started with hybrid pea seeds that were all round. These seeds then grew into hybrid plants. When these hybrid plants made seeds, Mendel's law was at work. One-fourth of the seeds were round. One-half were round hybrid seeds. One-fourth of the seeds were wrinkled. In his paper Mendel explained his results. He thought there were two kinds of pollen and egg cells.

Those who looked at Mendel's paper also thought about his results. They knew that Mendel's hybrid pea plants have a pair of forms, round and wrinkled. They saw how two kinds of pollen and egg cells could be produced. The round and wrinkled forms must split apart. Half the pollen would carry the round form. The other half would carry the wrinkled form. Likewise, half the egg cells would carry the round form. Half would carry the wrinkled form. So the 1 to 2 to 1 law became the Law of Segregation.

A few years after 1900, scientists knew something else about Mendel's hybrid pea plants. They knew that a pair of genes tells these pea plants to make round and wrinkled seeds. One gene says "make a round seed." The other gene says "make a wrinkled seed." So the Law of Segregation had to change. Today the law says that a pair of genes separates when pollen and egg cells are created. The pollen and egg cells each carry one member of the gene pair.

Genes are in pairs because plants and animals have their chromosomes in pairs. Peas have seven different chromosomes, and each one has a mate. In a pea plant with round hybrid seeds, one member of the chromosome #7 pair has the gene for round seeds. The other member has the gene for wrinkled.

Pollen grains get one member of each of the seven chromosome pairs. So do egg cells. In the hybrid pea plant, half the pollen grains get a chromosome #7 that has a round gene. The other half get a chromosome #7 that has a wrinkled gene. The same is true for the egg cells.

When an egg cell is fertilized, a new seed is created. It receives a complete set of chromosomes from the egg cell and a complete set from the pollen grain. The new seed then has a double set of chromosomes.

Scientists would look at Mendel's paper again. They saw Mendel's combination series law. This was his second law for hybrids. He had started with pea plants that made only round yellow seeds and plants that made only wrinkled green seeds. The hybrid seeds from these two plants were all round and yellow. Mendel then planted these hybrid seeds and tended the new plants. When these plants made seeds, Mendel's law was at work. He had four types of seeds: round yellow, round green, wrinkled yellow, and wrinkled green.

The scientists saw how all these combinations could turn up. For this to happen, all forms of the traits must be free to join each other. So the combination series law became the Law of

Independent Assortment. Today the law says that genes are inherited independently of each other. That's why the genes for seed shape and seed color can join in different combinations. These combinations produce four types of seeds.

This law doesn't work with all combinations of genes. For the law to work, the genes for each trait must be located on different chromosomes. Or they must be located far from each other on the same chromosome. If the genes are too close together, they are connected or linked. They are mostly inherited as a group. These genes are not likely to be inherited independently of each other. The result is that linked genes do not combine with other genes very often.

In 1900 Carl Correns found linkage in a plant called snowflake. In 1906 three British scientists found the second example of linkage. R. C. Punnett, E. R. Saunders, and William Bateson were working with sweet peas. (These are not the same plants Mendel used.) In sweet peas the genes for flower color and pollen shape are linked.

This is a simplified gene map of chromosome #1 in a garden pea. It shows the location of pairs of genes for two traits—seed coat color and seed color. An actual choromosome would have just one member of each gene pair.

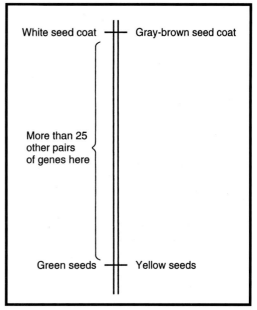

White seed coat — Gray-brown seed coat

More than 25 other pairs of genes here

Green seeds — Yellow seeds

The discovery of linkage turned out to be a good thing. It gave scientists a way to find the location of genes on a chromosome. A few years later, the first gene map of a living thing was created.

Genetics was just beginning to answer many kinds of questions. In the 1920s scientists joined genetics with the ideas of Charles Darwin. Darwin had explained how life evolves, or changes over time. But Darwin had not known about the laws of heredity. Now these laws improved his ideas. By the 1940s scientists saw how this combination of genetics and evolution could work. It was as if science had received a magical key that unlocked many secrets of life.

Today scientists are seeing evolution in action and learning more about it.

Genetics was ready to unlock more secrets of life. In 1953 James Watson and Francis Crick announced their discovery of the structure of DNA. This molecule has the instructions to make features or traits, written in special chemical codes. A chromosome contains one molecule of DNA. And the DNA, in turn, contains many genes.

Today scientists are learning more about the genes of all living things, including us. It is an exciting time. Many questions remain, and the search for answers goes on.

Like Mendel, scientists can study the seed shape and other traits of a pea plant. They are looking at the phenotype, or visible features, of the pea plant. Unlike Mendel, scientists can study the genes for seed shape and other traits. They are looking at the genotype, or kinds of genes, in the pea plant.

The seed-shape gene comes in two forms. One form contains the chemical code to make a

DEM NATURFORSCHER
P. GREGOR MENDEL
1822 – 1884

ERRICHTET 1910 VON FREUNDEN DER WISSENSCHAFT

The Gregor Mendel memorial in Brno honors his contribution to science. The wording (in German) says: To the investigator Father Gregor Mendel, 1822-1884, Erected in 1910 by the Friends of Science.

round seed, and the other contains the code to make a wrinkled seed. These two forms of the seed-shape gene are known as alleles.

Scientists now know why pea seeds are round or wrinkled. They are adding to the story that Mendel began.

Gregor Mendel started by asking questions about plant hybrids. His work with peas gave him some answers. Others rediscovered his work and looked at it in a different way. For them, Mendel's paper on peas answered questions about heredity. Gregor Mendel had become the father of genetics.

Mendel could not have known that this would happen. He would have been very surprised. Yet he knew all along that his work was important and the world would learn about it. For as he once said to his friend Gustav von Niessl: "My time will come."[1]

In Mendel's Footsteps

Gregor Mendel kept careful records of his experiments. This helped him see any interesting results. Yet even simple activities can give you interesting results. By writing down what you see, you have taken the first step in Mendel's footsteps.

Flipping a Coin

In his experiments, Mendel saw chance at work. In this activity, you can also see chance at work.

Let's say you have a hybrid pea plant. This pea plant has the gene for round seeds and the gene for wrinkled seeds. A pollen grain from this plant will have either a gene for round or a gene for wrinkled. Either gene is equally likely. What is the chance of picking a pollen grain with a round gene? This activity will help you find the answer.

Materials Needed

- A coin with a "head" and a "tail" (A wooden nickel will also work.)
- Paper
- Pencil

Procedure

1. Let a head stand for a round gene and a tail stand for a wrinkled gene.
2. Write Trial #1 on a sheet of paper.
3. Flip the coin and record if it is a head or tail.
4. Repeat step 3 nine times.
5. Write Trial #2 on a new sheet of paper.
6. Flip the coin and record the result.
7. Repeat step 6 thirty-nine times.
8. Figure out the fraction of heads in each trial. For example, in Trial #1 you might have recorded 6 heads out of 10 flips, or 6/10.

What was your fraction of heads in Trial #1? Did the fraction change in Trial #2?

Explanation

The fraction of heads will be close to one-half after you do many coin flips. That's another way of saying that the probability of a head is one-half. Now suppose you have hundreds of pollen grains instead of the coin flips. The probability of picking a pollen grain with a round gene is one-half.

Flipping Two Coins

In one experiment Mendel had hybrid seeds that were all round. Mendel planted the hybrid seeds, and the hybrid plants produced a new crop of seeds. That's when things began to get interesting. This activity explains why Mendel ended up with his famous ratio.

Materials Needed
- Two coins, each with a "head" and a "tail"
- Paper with lines
- Pencil

Procedure
1. Let one coin stand for a pollen grain and the other coin stand for an egg cell.

2. Write pollen on the upper left side of the paper and write egg cell on the upper right side.

3. Let a head stand for a round gene and a tail stand for a wrinkled gene.

4. Flip the "pollen" coin and record if it is a head or tail.

5. Flip the "egg cell" coin and record the result. Mark this on the same line as the result for the "pollen" coin. Then move down a line.

6. Repeat steps 4 and 5 nineteen times.

Did you get all four possible combinations? The four are head-head, head-tail, tail-head, and tail-tail. The head-tail and tail-head combinations are the two ways of coming up with a head and a tail. Now calculate the fraction of head-head combinations you had out of the number of times you did steps 4 and 5. You did these steps a total of twenty times. Next, calculate the fraction of head-tail and tail-head combinations out of twenty. Your numerator will be the sum of the head-tail and tail-head combinations. Lastly, calculate the fraction of tail-tail combinations out of twenty. Is one of your fractions greater than the other two?

Explanation

The head-head combination stands for two round genes. This round seed will grow into a plant that will make only round seeds. The tail-tail combination stands for two wrinkled genes. This wrinkled seed will grow into a plant that will make only wrinkled seeds. What about a head and a tail or vice versa? This combination of a round gene and a wrinkled gene is a hybrid seed. It is a round seed, since round is dominant over wrinkled. This round seed will grow into a plant that will make both round and wrinkled seeds.

If you repeat steps 4 and 5 many times, you would expect to get a fraction of one-fourth for the head-head combination. You would expect a fraction of one-half for the combination of head-tail and tail-head. You would expect a fraction of one-fourth for the tail-tail combination. So it makes sense that Mendel in his pea experiments would get a ratio of 1:2:1. You may have to do many coin flips to get close to this ratio. We know that Mendel had to grow many plants for his experiments. He wanted to be sure of his results.

Looking for Traits

Mendel spent time looking for traits in his pea plants. He decided to use seven traits in his experiments. He looked at two forms of each trait. For example, for the trait of seed shape, he looked at round and wrinkled seeds.

Mendel could have used other traits in his pea plants. In this activity, you can spot some of them by growing your own peas. You won't need to grow as many pea plants as Mendel did. Just a few will do.

Materials Needed

- Two packages of pea seeds (Each package should contain a different variety of pea seeds. Gardening stores or supermarkets carry many varieties to pick from.)
- 2 large plastic cups or small pots
- Potting soil or backyard soil
- Water
- Marking Pen
- Paper
- Pencil

Procedure

1. Add potting soil or backyard soil to the cups or pots.

2. Put 3 seeds from one package in a cup or pot. Cover with about one-half inch of soil. Save the leftover seeds.

3. Using a marking pen, label the cup or pot with the initial of the variety you are growing.

4. Add a small amount of water to the cup or pot.

5. Repeat steps 2–4 for the second package.

6. Place cups or pots near a sunny window or outside.

7. Add a small amount of water to each container every day.

8. Let the plants grow for a few weeks. Then look at the stems of the plants. Compare the two different varieties you have grown. Does one variety have green stems and the other have yellow stems? Does one variety have thicker stems than the other? Also try looking for slender curved things called tendrils at the ends of the leaves. Some peas have them and some don't.

9. Look for traits in the leftover seeds. Compare the two varieties. Do they

have purple spots or not? Are their surfaces indented or smooth?

10. Put the name of each variety on a separate sheet of paper. Do the varieties have the same forms of each trait? For example, do both have green stems? Write down what you see and look for differences.

Explanation

Two varieties of peas may have different forms of the same trait, depending on the genes they carry. You are looking for traits you can see easily. You may spot some other traits. You might try growing peas with pods you can eat. Also grow peas with pods you can't eat. That gives you another trait—pod edibility. The two forms are edible and non-edible. The non-edible pods contain a tough substance called parchment. Edible pods lack parchment.

Using pea plants, Gregor Mendel discovered how nature works. He carefully planned his experiments and recorded his results. Today there are still plenty of discoveries to be made. Maybe you will make one.

Chronology

1822—July 22: Johann (Gregor) was born to Rosine and Anton Mendel in Heinzendorf, Silesia.

1834–1840—Johann attended school in Troppau.

1840–1843—Johann studied at Olmütz Philosophical Institute.

1843—September 7: Johann accepted by Augustinian monastery of Saint Thomas in Brünn, Moravia; October 9: Johann took Gregor as his first name.

1845–1848—Gregor studied to become a priest at Brünn Theological College.

1847—August 6: Gregor became an ordained priest.

1849—Mendel became a teacher at Znaim.

1851–1853—Mendel studied science at the University of Vienna, Austria.

1854–1868—Mendel taught science at Brünn Modern School.

1856–1863—Mendel did experiments with pea plants.

1857—Gregor's father, Anton Mendel, died.

1862—Gregor's mother, Rosine Mendel, died.

1865—February and March: Mendel spoke about his experiments with pea plants at Brünn Natural Science Society.

1866—Mendel's paper titled "Experiments on Plant Hybrids" published by Brünn Natural Science Society.

1866–1871—Mendel did experiments with hawkweed and other plants.

1866–1873—Mendel wrote letters to Carl von Nägeli.

1868—March 30: Mendel elected abbot of Augustinian monastery of St. Thomas.

1869—June: Mendel spoke about his experiments with hawkweed at Brünn Natural Science Society.

1871—Mendel's paper titled "The Whirlwind of October 13, 1870" published by Brünn Natural Science Society.

1884—January 6: Gregor Johann Mendel died at Augustinian monastery of St. Thomas.

1900—Mendel's work on plant hybrids rediscovered by Hugo de Vries, Carl Correns, and Erich von Tschermak. The science of genetics was born.

Chapter Notes

Chapter 1

1. Vítězslav Orel, *Mendel* (New York: Oxford University Press, 1984), pp. 1–2.

Chapter 2

1. This duty to work for the nobleman was known as the corvée.

2. Hugo Iltis, *Life of Mendel* (New York: W. W. Norton and Company, 1932), p. 37.

3. Ibid., p. 38.

Chapter 3

1. Hugo Iltis, *Life of Mendel* (New York: W. W. Norton and Company, 1932), p. 42.

2. Ibid., p. 51.

Chapter 4

1. Hugo Iltis, *Life of Mendel* (New York: W. W. Norton and Company, 1932), p. 62.

2. Ibid., p. 81.

Chapter 5

1. Vítězslav Orel, *Gregor Mendel: The First Geneticist* (New York: Oxford University Press, 1996), p. 122.

2. Hugo Iltis, *Life of Mendel* (New York: W. W. Norton and Company, 1932), p. 108.

3. Ibid.

4. Ibid., p. 97.

Chapter 6

1. Alain Corcos and Floyd Monaghan, *Gregor Mendel's Experiments on Plant Hybrids: A Guide of Study* (New Brunswick, N.J.: Rutgers University Press, 1993), p. 101.

Chapter 7

1. Hugo Iltis, *Life of Mendel* (New York: W. W. Norton and Company, 1932), p. 99.

2. Curt Stern and Eva Sherwood, eds., *The Origin of Genetics: A Mendel Source Book* (San Francisco: W. H. Freeman and Company, 1966). Part one contains the letters Mendel sent to Carl von Nägeli.

3. Iltis, p. 155.

4. Vítězslav Orel, *Gregor Mendel: The First Geneticist* (New York: Oxford University Press, 1996), pp. 131–156.

Chapter 8

1. Hugo Iltis, *Life of Mendel* (New York: W. W. Norton and Company, 1932), p. 240.

2. Ibid., p. 245.

Chapter 9

1. Robert Davies-Jones, "Tornadoes," *Scientific American*, August 1995, pp. 48–57.

2. Hugo Iltis, *Life of Mendel* (New York: W. W. Norton and Company, 1932), p. 229.

Chapter 10

1. Sr. C. W. Eichling, "I Talked with Mendel," *Journal of Heredity*, July 1942, p. 245.

Chapter 11

1. Hugo Iltis, *Life of Mendel* (New York: W. W. Norton and Company, 1932), p. 282.

Glossary

allele—A form of a gene. In peas, the gene for seed shape has two alleles—round and wrinkled.

breeding—Producing plants and animals with valuable characteristics.

chromosome—Rod-shaped structure in a cell. It carries genes, which contain the codes for features or traits.

DNA—A molecule containing chemical codes to make features or traits in a living thing. Genes are tiny sections of DNA.

dominant—A form of a gene that masks another form of the same gene. In peas, the round form of the seed-shape gene is dominant—it masks the wrinkled form.

gene—Material containing a chemical code to make a feature or trait, such as seed shape in peas. Genes are found on the chromosomes.

genetics—The science that studies heredity.

genotype—The kinds of genes in a living thing. For example, a pea plant that makes both round and wrinkled seeds has one gene for round seeds and one gene for wrinkled. A pea plant that makes only wrinkled seeds has two genes for wrinkled seeds.

heredity—The passing of features or traits from one generation to another.

hybrid—The offspring of two parent plants that have different genes. One parent may be a pea plant that makes only tall plants. The other parent may be a pea plant that makes only short plants. The two parent plants produce offspring that are all tall.

Law of Independent Assortment—A rule of heredity stating that genes are inherited independently of each other. To see the law at work, start with pea plants that make only round yellow seeds and pea plants that make only wrinkled green seeds. The offspring are hybrid seeds that are all round and yellow. Now allow the hybrid seeds to grow into hybrid plants. These plants will make four types of seeds: round yellow, round green, wrinkled yellow, and wrinkled green. The genes for seed shape and seed color are inherited in different combinations. These combinations produce four types of seeds. This law doesn't work with all combinations of genes. For the law to work, the genes for each trait must be located on different chromosomes. (The genes for seed shape and seed color are.) The law also works if the genes are located far from each other on the same chromosome. If the genes are too close together, they will mostly stay connected, or linked, when passed to offspring. See also Linkage.

Law of Segregation—A rule of heredity stating that a pair of genes separates when pollen and egg cells form. The pollen and egg cells each carry one member of the gene pair. A pea plant that makes round and wrinkled seeds has a pair of genes for these two shapes. One gene has the code to make round seeds, and the other gene has the code to make wrinkled. Half of the pollen will carry the gene for round seeds, and half will carry the gene for wrinkled seeds. Half of the egg cells will carry the round gene, and half will carry the wrinkled gene. Genes are in pairs because plants and animals have their chromosomes in pairs.

linkage—A connection of genes located close together on the same chromosome. These genes are mostly inherited as a group. They are not likely to be inherited independently of each other. The result is that some combinations of genes will not turn up as often as others. See also Law of Independent Assortment.

phenotype—The visible features of a living thing. For a pea plant this includes seed shape, plant height, and many other features.

recessive—A form of a gene that is masked by another form of the same gene. In peas, the wrinkled form of the seed-shape gene is recessive—it is masked by the round form.

Further Reading

Aronson, Billy. *They Came From DNA*. New York: W. H. Freeman and Company, 1993.

Asimov, Isaac. *How Did We Find Out About Genes?* New York: Walker and Company, 1983.

Balkwill, Fran. *DNA Is Here to Stay*. Minneapolis: Carolrhoda Books, 1993.

———. *Amazing Schemes Within Your Genes*. London: Collins, 1993.

Corcos, Alain, and Monaghan, Floyd. *Gregor Mendel's Experiments on Plant Hybrids: A Guided Study*. New Brunswick, NJ: Rutgers University Press, 1993.

Hershey, David. *Plant Biology Science Projects*. New York: John Wiley & Sons, 1995.

Index